contents

solid as a rock ·· 4
beginning of a revolution ······································· 6
from core to crust ··· 14
probing the ocean depths ····································· 21

shifting mosaic ·· 28
understanding plate tectonics ···························· 30
recent discoveries ·· 40

danger zones ·· 44
earthquake tremors ··· 46
volcanic activity ··· 51
living with danger ·· 60

glossary ··· 66
index ·· 68
further reading/acknowledgments ······················ 72

solid as a rock

U ntil the 1960s, the map of the Earth was considered static. Geologists knew that there had been upheavals in the hundreds of millions of years of the Earth's past, but received scientific wisdom was that the position of the world's land masses and oceans was fixed and durable – solid as a rock, in fact. These ideas held sway because the alternative – that we live on a constantly shifting mosaic of continents, surrounded by transient oceans – was literally unbelievable. A revolution in understanding built up over time as a result of an explosion in the information available to scientists. The use of dating techniques based on radioactivity allowed them to measure the immense age of the Earth, providing a large enough timespan for previously unimaginable changes. A new understanding of the Earth's magnetic field opened up fresh perspectives, while the observation of earthquakes gave insight into the structure of the hidden depths of the planet. Finally, it was the exploration of the ocean floor that revealed the key to the mystery of the restless Earth.

himalayan peak
It may look solid and immovable, but, in fact, this mountain, like the whole Himalayan mountain chain, was once low-lying continental crust. Between 10 and 20 million years ago the Indian subcontinent collided with the continent of Asia, causing the crust to buckle, fold, and rise under the immense force of the impact. The Himalayas are the world's highest mountain chain.

beginning of a revolution

Modern understanding of the history of the Earth got under way in the late 18th century, when individuals, such as the Scottish geologist James Hutton (1726–97), established the basic principle that the different rock strata seen as layers in cliffs and gorges were a visible record of past time. Hutton also established the maxim of "uniformitarianism", which stated that the same processes that shape the landscape around us in the present caused the features we can observe in rocks from the distant past. In his immensely influential book *Principles of Geology*, written in 1830, British geologist Sir Charles Lyell confirmed this principle, insisting that even the most dramatic geological features could be accounted for by very gradual changes over immense periods of time – by the action of waves and weather, occasional volcanic eruptions, and so forth. Lyell's work influenced that of biologist Charles Darwin (1809–82), whose *On the Origin of Species,* published in 1859, postulated that, over long periods, minute changes in each generation of creatures had allowed the evolution of the animals that exist today.

One crucial problem that confronted exponents of both Lyell's view of geology and Darwinian evolution was the

Sir Charles Lyell (1797–1875), a Scot who read law at Oxford University, made extensive observations of present-day processes at the surface of the Earth. He established not only that these processes could have formed rock in the past, but also that the processes would take hundreds of thousands of years. The concept had been set out by James Hutton 40 years before, but Lyell's books brought it into the scientific mainstream.

immense amount of time that it must have taken for the subtle processes responsible for change to occur. Some scientists felt that the Earth could not be anything like old enough for these processes to happen.

For example, the physicist William Thomson (1824–1907), later Lord Kelvin, came up with the figure of 40 million years as the maximum time available for the evolution of the Earth. Basing his work on measurements of the temperature in mines, he calculated the Earth must have taken 20 to 40 million years to cool down from the ball of molten rock that he correctly assumed it originally had been. Although Kelvin's calculations made the Earth considerably older than had traditionally been believed, his estimate left wholly inadequate time for biological evolution or for the slow accumulation of sand and mud on the Earth to build up into the vast thicknesses of rock that make landmasses and mountains.

molten earth
About 4.6 billion years ago, a red-hot ball of molten matter emerged from a huge condensing cloud of gas and dust that rotated around a newly formed Sun. The Earth was born.

dating the earth

In fact, Kelvin's calculation of the age of the Earth was wildly mistaken, chiefly because he did not know about radioactivity. Radioactive processes inside the Earth generate enormous amounts of heat, so preventing the planet from cooling as rapidly as Kelvin had supposed.

Ironically, radioactivity also provided the key to the accurate measurement of the age of the Earth which had eluded Kelvin. Early 20th-century scientists explored the use of rates of radioactive decay (see p.39) as a means of dating rocks. Early radiometric dating was not especially precise, but it pushed the age of the Earth firmly into the many hundreds of millions of years needed to allow the

the history of the earth

Most of the world's land surface is covered with sedimentary rocks – layers of debris deposited over millions of years and compressed to form rock strata. In the 17th century, Danish naturalist Nicolaus Steno asserted that the youngest strata were always at the top and the oldest at the bottom. This principle is in fact true only where strata have not been disturbed by later upheavals. By observing what fossils are found in different strata, however, geologists have been able to work out the sequence of rock formations even where strata are buckled or twisted. The order of strata and the fossil record tell us the relative age of rocks – which are younger and which older – but not, in absolute terms, how old they are.

geological time chart

Observing strata and fossils around the world allowed geologists to divide up the history of the Earth into a sequence of named periods. The age of the rocks in these periods was established later by other dating methods (see p.39).

key

mudstone   sandstone

limestone   shale

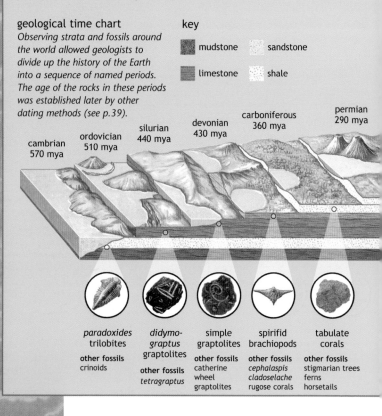

cambrian
570 mya

ordovician
510 mya

silurian
440 mya

devonian
430 mya

carboniferous
360 mya

permian
290 mya

paradoxides
trilobites

other fossils
crinoids

*didymo-
graptus*
graptolites

other fossils
tetragraptus

simple
graptolites

other fossils
catherine
wheel
graptolites

spirifid
brachiopods

other fossils
cephalaspis
cladoselache
rugose corals

tabulate
corals

other fossils
stigmarian trees
ferns
horsetails

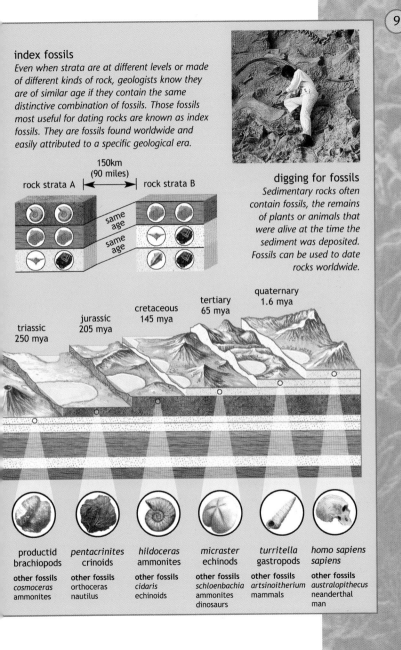

index fossils
Even when strata are at different levels or made of different kinds of rock, geologists know they are of similar age if they contain the same distinctive combination of fossils. Those fossils most useful for dating rocks are known as index fossils. They are fossils found worldwide and easily attributed to a specific geological era.

150km (90 miles)

rock strata A

rock strata B

same age

same age

digging for fossils
Sedimentary rocks often contain fossils, the remains of plants or animals that were alive at the time the sediment was deposited. Fossils can be used to date rocks worldwide.

quaternary 1.6 mya

tertiary 65 mya

cretaceous 145 mya

jurassic 205 mya

triassic 250 mya

productid brachiopods
other fossils
cosmoceras ammonites

pentacrinites crinoids
other fossils
orthoceras nautilus

hildoceras ammonites
other fossils
cidaris echinoids

micraster echinods
other fossils
schloenbachia ammonites dinosaurs

turritella gastropods
other fossils
artsinoitherium mammals

homo sapiens sapiens
other fossils
australopithecus neanderthal man

steady accumulation of sedimentary rocks from the sloppy sand and mud found at the surface. The age of the Earth is now estimated at around 4.6 billion years.

The timespans revealed by radiometric dating seemed to confirm the gradualist view of the formation of rocks and landscapes, but anomalies remained, including problems thrown up by the study of fossils. Puzzlingly, the coal seams of Britain had been found to be full of the relics of tropical plants. Dinosaur fossils had also been found in the British Isles, yet surely these creatures must have lived in a warm climate, like today's large reptiles? There was other evidence that the climate in the past had been very different from that of the present. For example, rocks carved by great sheets of ice were found in places such as South Africa, far from the poles. There had clearly been plenty of time for different areas of the Earth to have undergone dramatic changes in climate, but why and how this had happened was unclear.

continental drift

One of the explanations of the diverse conditions that appeared to have existed in the geological past proved to be very much ahead of its time. Alfred Wegener (1880–1930), a German astronomer and meteorologist, outlined the concept of "continental drift" in 1915. He drew together strands of evidence from geology, geography, and meteorology to suggest that the continents had moved across the face of the Earth. He asserted that the fossils of tropical plants found in the coals of northern Europe proved that Europe had, in the past, been closer to the equator. Similarly, ice scratches in rocks near the equator showed that these rocks had once been in a polar region.

fern evidence
The Glossopteris fern (above) has been found as a fossil (right) in coal seams on several continents, suggesting that those continents were once part of a single landmass.

shifting continents

In 1915 Alfred Wegener (1880–1930) suggested that the continents were once joined together to form one supercontinent, which he called Pangaea. Over millions of years, this single landmass split up into several smaller ones, which slowly drifted across the Earth's surface. His theory explained why certain groups of closely related animals and plants were found in widely separated countries across the globe.

alfred wegener

300 million years ago
Three continents came together to form one giant supercontinent.

pangaea

135 million years ago
Pangaea separated into two landmasses: Laurasia and Gondwana.

laurasia

gondwana

north america

eurasia

india

65 million years ago
Laurasia split into North America and Eurasia, and South America moved away from South Africa.

south america

south africa

virginia opossum
Marsupials exist in the Americas and Australia – both once joined to Antarctica.

india

eurasia

today
India is fused with Eurasia, and Australia is separate from Antarctica.

india

australia

antarctica

Wegener also looked at the shapes of the continents and suggested that the east coast of South America matched the shape of the west coast of Africa so well because they had once been joined together in a single vast continent. This would also explain the existence of similar rock formations in these two now far distant coastal areas.

Although continental drift did offer a potential explanation for much of the emerging information about the geological history of the planet, Wegener's theory was widely ridiculed. The very idea of continents moving about the Earth was considered preposterous and unscientific. The huge mass of circumstantial evidence Wegener had accumulated was considered inconclusive, especially in the face of the geologists' tried-and-tested maxim of uniformitarianism: that the present is the key to the past. If continents moved around the globe in the manner Wegener suggested, why was there no sign of their doing so now? And, above all, how could continents move through the solid rock of the ocean floor?

a confusion of theories

Wegener's notion of continental drift joined many other wild-sounding theories that were around at the same time, as scientists, eminent and otherwise, struggled to get to grips with how the Earth had changed through time. Ideas put forward included the suggestion that the Moon had been thrown off the Earth in a cataclysmic upheaval, leaving the Pacific Ocean as a scar behind it, and the proposal that the Earth had flipped over. Continental drift was seen as less respectable than such ideas (now known to be erroneous), above all because of its lack of any half-way plausible mechanism that would allow the continents to move.

To make progress with understanding and interpreting the evidence that was accumulating from the fledgling radiometric dating of rocks, from increasingly accurate geological mapping of the world, and from the beginnings of studies of the ocean beds, geologists needed to discover what lay beneath the Earth's surface. Until the 20th century, the interior of the Earth was a mystery. Apart from what could be seen in mines, and relatively shallow ones at that, there was simply no information. But all this would change with the arrival of new techniques – in particular, those used to study earthquakes – that uncovered the interior structure of the planet.

" I came to the idea on the grounds of the matching coastlines, but the proof must come from geological observations. "

Alfred Wegener, 1911

flipping earth
One wild theory proposed during the early years of the last century was that the whole planet had flipped over from north to south several times.

from core to crust

Although it is impossible to observe directly the interior of the Earth, geophysicists have discovered various ways of finding out about the planet's inner composition. One especially productive line of investigation has been the study of earthquakes, the tremors that result from sudden movements on faults underground (see p.48). The vibrations in the rocks around the source of the earthquake spread out like ripples when a pebble is thrown into a pool. These vibrations, known as seismic waves, can be measured by instruments called seismometers. Because the nature of the rock they travel through affects the seismic waves – for example, denser rock generally speeds them up – they can give information about the composition of the Earth between the earthquake source and the seismometer.

Geophysicists began to set up seismic observatories worldwide in the early years of the 20th century. In 1909 a Croatian seismologist, Andrija Mohorovičič, deduced from observation of two sets of seismic waves from a local earthquake that there was a discontinuity, or boundary, some 35km (21 miles) beneath the Earth's surface. He concluded that this was the boundary between the crust, the outermost layer of the Earth on which we live, and a denser mantle beneath. Observations worldwide subsequently confirmed Mohorovičič's findings. Further study of seismic waves uncovered other discontinuities deeper inside the Earth. Far from being a uniform ball of rock as originally supposed, the Earth was revealed as layered from its outer crust to its dense core.

seismic ripples
An earthquake creates concentric rings of vibrations through the Earth in the same way as a pebble creates concentric ripples when it is thrown into water.

the structure of the earth

Planet Earth is made up of three concentric layers: the crust, the mantle, and the core. These layers have been detected through the different effect each has on the seismic waves spreading from earthquakes. The Earth becomes hotter towards the centre, so much of its interior is partially or wholly molten. The exception is the inner core, which is under such great pressure that it is solid.

crust

mantle

outer core

inner core

lithosphere

layered earth

The crust and the upper part of the denser mantle beneath form a rigid shell around the Earth, known as the lithosphere. Between 100km (60 miles) and 300km (180 miles) thick, the lithosphere floats on a soft layer of partially molten mantle beneath. The outer core consists of molten iron, surrounding the solid inner core.

dominant rocks and minerals

On the crust, the continents are mostly made of granite and other crystalline rocks, while the ocean floors are chiefly basalt. The mantle is made of denser rocks such as peridotite, which is sometimes thrown to the surface by volcanoes. The core is mostly iron.

basalt

granite

peridotite

iron

crust composition

One interesting discovery about the interior of the Earth
was that the mantle below the lithosphere was soft and
able to flow slowly. The Earth's crust is floating on the
dense, fluid mantle. This finding helped explain why the
continents are so high and the ocean floors so low.

The continents are made of many types of rock, but their
foundations are essentially granite, a rock that was
originally molten, which then cooled to form an igneous
rock made of crystals. In most cases, the granite has been
heated and distorted under pressure, so that the minerals
are streaked out into bands, making the metamorphic rock
gneiss. Granite and gneiss contain a lot of the mineral
quartz, making them less dense than many rocks. In
contrast, rocks dredged from the ocean depths were found
to be almost entirely a tough, black, igneous rock called
basalt. This is the same rock that erupts from volcanoes in
the form of lava (molten rock). Basalt is denser than granite,
so oceanic crust floats lower than continental crust.

constant movement

The fluid layer in the mantle also allowed for the up and
down movements of the continents. Geologists could see
that continents might have risen and fallen because, for
example, fossils of sea creatures are common on land,

uplift and erosion
Over the last 60 million years, the Colorado Plateau has been gradually uplifted 3,000m (10,000ft). The river that runs through it has eroded the thick sedimentary layers to create the Grand Canyon.

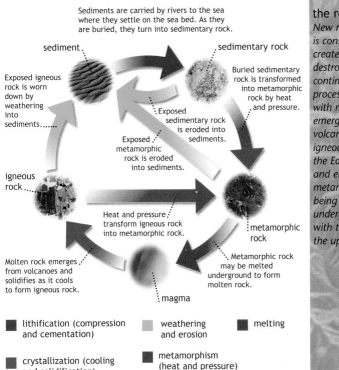

Sediments are carried by rivers to the sea where they settle on the sea bed. As they are buried, they turn into sedimentary rock.

sediment

sedimentary rock

Exposed igneous rock is worn down by weathering into sediments......

Buried sedimentary rock is transformed into metamorphic rock by heat and pressure.

Exposed sedimentary rock is eroded into sediments.

Exposed metamorphic rock is eroded into sediments.

igneous rock

Heat and pressure transform igneous rock into metamorphic rock.

metamorphic rock

Molten rock emerges from volcanoes and solidifies as it cools to form igneous rock.

Metamorphic rock may be melted underground to form molten rock.

magma

■ lithification (compression and cementation)

weathering and erosion

■ melting

■ crystallization (cooling and solidification)

■ metamorphism (heat and pressure)

the rock cycle
New rock material is constantly being created and destroyed in a continuous recyling process. It begins with magma emerging from volcanoes to form igneous rock on the Earth's surface and ends with metamorphic rock being melted underground to mix with the magma in the upper mantle.

even among the highest mountains. In fact, the continental crust is floating on the fluid mantle like a boat at sea. Just as taller boats have hulls that go further below the water, mountains have roots. So the higher, and therefore heavier, the mountain, the greater the amount of crust below it, extending into the mantle. Then, as the mountain is gradually eroded, the crust beneath floats upwards to maintain the equilibrium. So, rocks that once were at sea level could eventually be on mountain tops. Movement below and above the Earth's surface was also vital for the formation of sedimentary rocks – just one stage in the rock cycle (see above).

magnetic field

Although scientists were happy with the idea of uplift and subsidence, into the 1950s they remained convinced that there was no lateral drifting of continental landmasses. Yet new information from the study of palaeomagnetism – the history of the Earth's magnetic field – suggested otherwise. It was found that when rocks were formed or solidified, minerals containing iron were magnetized.

earth's magnetic field

The Earth is like a giant magnet. The magnetism is caused by the constantly moving liquid iron in the Earth's outer core. The magnetic field is so powerful that it affects every magnet on the planet. It also spreads out into space in the magnetosphere.

earth

the magnetosphere
Stretching 60,000km (37,000 miles) into space, the magneto-sphere protects us against charged particles that stream out from the Sun, known as the solar wind.

reversing polarization
From time to time the Earth's magnetic field reverses itself, so that compass north becomes south. By studying magnetized particles in rocks, geologists have been able to chart when the field was normal and when reversed back through millions of years.

magnetosphere creates shield against solar wind

van allen belt some charged

TIMELINE

| 0 | 10 | 20 | 30 | 40 | 50 |

(mya)   ☐ reversed   ■ normal

Using magnetometers, geophysicists were able to observe this trapped magnetism, which provided a record of the orientation and direction of the magnetic field at the time the rocks formed. It turned out that the Earth's magnetic field was surprisingly variable. In some cases, the magnetism trapped in the rocks was in completely the opposite direction to that of the magnetic field now. By correlating the magnetic readings with radiometric dating

northern lights
Some charged particles from the solar wind penetrate gaps in the magnetosphere near the magnetic poles, where they trigger these spectacular light shows.

lines of force
The Earth's magnetic field is shaped like that of a bar magnet; field lines run downward near the poles and along the surface at the Equator.

magnetosphere
elongated by
solar wind

pole   pole

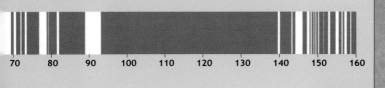

70   80   90   100   110   120   130   140   150   160

key points

• New instruments gave geophysicists their first glimpses inside the Earth.
• Some layers within the Earth were discovered to be fluid, not rigid.
• There were signs that the Earth was more changeable than scientists had thought: the magnetic field had reversed its polarity, not once, but many times.

of the same rocks, it became clear that in the past the magnetic field had reversed itself many times. This phenomenon became an aid in plotting the Earth's past, as geologists began to date rocks by their magnetic orientation as well as by radiometric methods.

Other anomalies in the magnetic record posed even more puzzling questions. Geophysicists represent the Earth's magnetic field by field lines, which show the orientation of a compass needle everywhere over the Earth. And because the field varies in three dimensions, field lines near the poles plunge steeply into the Earth, whereas those near the equator run parallel to the ground. Many palaeomagnetic measurements from lava flows showed ancient magnetic directions very different from the present field at the same location. Steep magnetic field readings near the equator, or flat-lying ones at high latitudes, suggested either that the rocks had moved considerable distances north or south around the world, or that the magnetic poles had moved.

change from the ocean

Despite all these exciting and challenging discoveries, until the 1960s geologists continued to work within the framework of a fixed map of the planet. There was, as yet, no sufficient evidence that the continents moved relative to one another, and no accepted mechanism that might make them able to move. The breakthrough was to come from our expanding knowledge of the ocean floor.

probing the ocean depths

The first clues about the relationship between the oceans and the continents came from discovering the varying depths, or shape, of the ocean floors. Until the late 19th century, the depths of the oceans were as much of a mystery as the interior of the Earth. From 1872 to 1876, HMS *Challenger* of the British Royal Navy made a pioneering attempt to map the ocean floors. The methods used on the *Challenger* expedition were fairly primitive – the vessel stopped approximately every 160km (100 miles) to pay out a weight on a line until it hit the ocean floor.

Traversing the Atlantic, the expedition found that the shallow seas near the coast first deepened, then, surprisingly, became shallower again, reaching a depth of only around 5km (3 miles), in an area now known as the Mid-Atlantic Ridge. The *Challenger* expedition's achievements included the discovery of the deepest part of the Pacific Ocean floor, 11.6km (7 miles) below the surface in the Marianas

challenging maritime quest
The four-year voyage of the Challenger *posed more questions than it answered about the deep ocean floor, finding relatively shallow water where only deep was expected, and extreme depths in narrow trenches close to the Pacific Ocean shore.*

Trench, an area that is now called the Challenger Deep. But the *Challenger* could provide little evidence of the shape or extent of the mid-oceanic underwater ridge that it had also discovered. It would be nearly another century before the East Pacific Rise was properly mapped, and mid-ocean ridges became recognized as one of the fundamental features of the ocean floor.

mapping the ocean floors

From the 1920s, many more depth measurements were taken as a result of the increasingly extensive laying of communications cables. It was established that the unexpectedly shallow region running down the middle of the Atlantic Ocean was an undersea mountain range. Other expeditions were launched in the 1920s and 1930s,

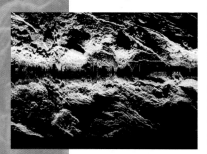

notably that of the *Dana*, which found a similar ridge in the Indian Ocean and named it the Carlsberg Ridge.

By the 1950s, the use of echo sounders had given rise to a new level of precision in the mapping of the ocean floors. The vague picture of some sort of craggy shallow area in the mid-Atlantic was replaced by an image of a well-defined ridge running throughout the ocean. The ridge had a rift valley along its centre, and on each side of the ridge the ocean floor gradually deepened away.

undersea mountain range
This sonar image of the Mid-Atlantic Ridge reveals a mountain range cut along the centre by a rift valley. Each side slopes away to the deep ocean floor, falling some 2km (1.25 miles) from the peaks.

Meanwhile, close to some of the edges of the oceans, notably the margins of the Pacific, the very deepest parts of the seas formed long, narrow trenches. These trenches were not generally located far out in the ocean, but were often close to land, running parallel to the shore, or, intriguingly, parallel to lines of volcanic islands, especially around the rim of the Pacific.

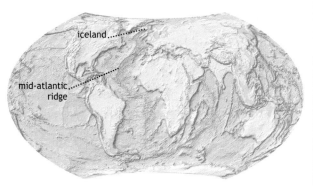

ocean ridges
The submarine mountains found in every ocean are depicted in this relief map that also shows the mountain ranges on land. The Mid-Atlantic Ridge appears above sea level in Iceland.

dating the ocean floors

The biggest step forward in the understanding of the relationship between continents and oceans came from discovering the age of the ocean floors. In the 1930s and 1940s, radiometric dating of rocks from the continents had pushed the age of the Earth's crust back to around 3,000 million years. But when dredged samples of rocks from the ocean floors were dated, it appeared that none of them was more than 200 million years old, and many were much younger. The oceans showed an approximate pattern of older rocks nearer the edges and younger ones at the centre. Those from the ridge at the centre of the oceans were the youngest of all.

As early as 1931 English geologist Arthur Holmes (1890–1965), a pioneer in the radiometric dating of rocks, had suggested that the mantle beneath the oceans might be flowing, carrying the continents apart. At that time his tentative proposals, which provided a possible mechanism for Wegener's continental drift, had not been taken

When the British geophysicist **Arthur Holmes** (1890–1965) was born, the Earth was generally thought to be around 6,000 years old; when he died, it was known to be around 4.6 billion years old. Holmes took the lead in extending the probable age of the Earth by applying emerging ideas about radioactivity to the geological time scale, which was then solely dependent on a relative dating regime.

During the Second World War, American geologist **Harry Hess** (1906–69) mapped the topography of the ocean floor from submarines. He was surprised by the amount of heat flowing from the ocean floor. In a paper published in 1962, he drew together ideas from different branches of the science – ocean mapping, earthquake location, gravity measurements, composition, and the age of the ocean floor – and came up with the model of the sea floor spreading that we have today.

seriously. But, as the dates began to mount up, always in the same pattern, it began to look as if he had been on to something.

deep-ocean islands

Further evidence supporting the observation that the youngest crust ran along the middle of the ocean floor came from investigation of ocean islands. Unlike the lines of islands along the edges of oceans, or in, for example, the Caribbean, these are single islands or chains of islands out in the deep ocean. Originally volcanoes, they are now often swathed in limestone grown by coral reefs. Dating of their rocks showed that the youngest islands were invariably closest to the middle of the oceans, and the oldest furthest away.

The pattern was reinforced by the discovery of coral atolls and seamounts, often in line with chains of ocean islands. Seamounts are volcanoes that once poked above the sea to make islands but are now submerged. Atolls, such as Bikini atoll, famous for being used in nuclear tests, are ring-shaped coral reefs formed above such sunken volcanic islands. Again, the older the seamount, the further it is from the ridge. The older islands and seamounts were also in deeper ocean. These features all supported the idea of the ocean crust somehow forming at the ridges along the centre of the oceans. They also indicated that the ocean floor sank as it became older.

Supporting evidence came from the magnetism locked in the basalt rocks of the sea floor when they formed. The palaeomagnetic data from oceanic rocks was far less confusing than that from continental rocks, which gave

recap

Earth's **magnetic field** arises in the molten iron of the core. It has changed polarity many times over the Earth's history. The ages of rocks showing normal and reversed polarity combine to give a magnetic timescale, which confirmed the age and origin of the sea floor.

evidence of many complex shifts. Some readings taken from the sea floor were very like the Earth's magnetic field today; others were the complete opposite. In fact, they were recording periods of normal and reversed magnetic field from the Earth's past.

how the oceans form

In 1963, English geophysicists Frederick Vine (1939–88) and Drummond Matthews (1931–97) took the idea of the sea floor ageing from the centre outwards and added to it the idea of magnetic reversals, to put forward a testable model for how the oceans form. They suggested that new ocean floor was constantly being created along the line of the rift valley at the centre of the volcanic mid-ocean ridges, by magma rising from the Earth's mantle. This new crust was then carried to the sides of the ocean as if it were on a conveyor belt.

Their theory predicted that, when more exact magnetic maps of the ocean floors existed, they would show a symmetrical pattern of magnetic stripes on the central ridge. Whenever the magnetic field made its mysterious reversals, the pattern would have been fixed in the ocean floor basalt as it solidified and then split apart. Better magnetic maps showed exactly the pattern predicted – stripes of rocks magnetized in opposite directions, running up and down the middle of the oceans, roughly parallel to the ridges in the centre.

plunged into the abyss
Oceanic and continental crust moves inexorably to the edge of the conveyor belt where it dives into the intense heat of the mantle to become molten rock.

convection in the mantle

What was the force behind the conveyor belt? If the Earth was moving, there had to be something driving it. Ocean crust forming in the middle and moving outwards harked back to suggestions of some sort of convection in the mantle of the Earth, below the crust. The formation of new ocean crust seemed to be evidence for the upward flow of hotter material along the centre of the oceans, leading to molten rock erupting and solidifying. The steady movement of the ocean floor away from the centre, it was argued, formed the upper part of a convection cell, heated from the core below.

> ❝ It will be difficult for most of us to accept that large amounts of what we have written and taught has been erroneous. ❞
>
> John Tuzo Wilson, Canadian geophysicist, 1963

If upward convection was occurring in the mantle, then there must be regions where material moved downwards, too. The obvious place to look was the edges of the oceans. It appeared that the deepest parts of the oceans, the trenches, lay above the place where the ocean floor bent down to start its journey back into the mantle – the downward part of the huge convection cell – completing the circuit. This explained why there were no really old rocks on the ocean floor: they were constantly recycled back into the mantle.

mechanism for movement

By the late 1960s, the great objection to continental drift – how could the continents plough through the solid rock of the ocean floors – had vanished. The ancient continents and the young oceans complemented each other. Ocean crust had an inherently shorter lifespan than continental crust as it formed, cooled, and slid beneath the more buoyant continents. As oceans split apart and spread they could carry continents with them.

key points

• The world's oceans have ranges of undersea mountains along their centres.

• Ocean floor forms at these mid-ocean ridges, and becomes older the further it is away from the ridge.

• Circulation in the mantle carries the ocean floor away from the ridge, as if on a conveyor belt.

convection cell

Scientists believe that convection cells in the Earth's mantle may be the huge conveyor belts that cause oceanic and continental crusts to move.

Convection is the transmission of heat through a liquid or gas by means of circulating currents. It happens when you heat up a pan of water on the stove. Parts of the water near the bottom of the pan warm up, and become less dense, so they move upwards, displacing cooler (and therefore denser) water downwards. If the zones of upward and downward flow join together to form loops, this is known as a convection cell. The enormous temperatures deep in the mantle may make rock convect upwards and then loop down again as it cools. A mantle convection cell would take millions of years to complete one cycle.

the effect of convection
Sections of the Earth's crust may move because they are carried along by convection currents in the mantle on which they float.

Not all in the earth sciences welcomed these revolutionary ideas with open arms. There were plenty of established figures who maintained that the continents were fixed and always had been. There were tremendous rows within institutions, at conferences, and in the scientific press. Established researchers had to examine completely new lines of evidence, change the way they had looked at the world, and, often, change their minds. But slowly, the weight of evidence brought most people round. Research began to look in earnest at the new theory of continental drift and its implications – and at the many gaps that still remained in this overarching idea.

shifting mosaic

S ince the 1960s, the theory of plate tectonics has provided scientists with a different way of thinking about the Earth. The new model sees the Earth's surface as divided into rigid plates that float on the fluid mantle beneath. The continents and oceans move as blocks, sometimes splitting or joining together. Oceans are created and destroyed as part of the process, their waxing and waning allowing the continents to migrate. The movement is continual, if almost imperceptibly slow, and is concentrated at the boundaries between plates, where we experience it as earthquakes. Where continents collide, great mountain ranges have formed. The drifting of continents that has endlessly reshaped the map of the world can explain many long-standing mysteries, such as why the fossils of tropical plants are found in cool northern lands, and why volcanoes occur where they do. (The pattern of distribution of volcanoes around the world reflects the interruption to the Earth's layered structure at the boundaries of tectonic plates.)

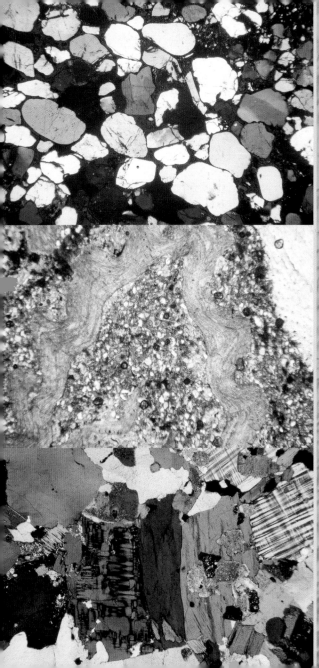

multi-layered crust

Thin layers of sedimentary rocks, such as sandstone (top) and limestone (centre), cover 75 per cent of the Earth's land surface. These layers are continuously deposited over igneous rock, such as granite (bottom), and metamorphic rock, such as gneiss.

understanding plate tectonics

It is now understood that the Earth's shell, the lithosphere, which consists of the crust and the rigid upper layer of the mantle, floats on the soft, semi-molten lower layer of the mantle, the asthenosphere. The lithosphere is split into nine large slabs and a dozen smaller ones. These pieces of the Earth's shell are known as tectonic, or "building" plates. The theory of plate tectonics assumes that the entire surface of the Earth is a mosaic of plates, without gaps, and that these great plates, both continent and ocean, are moving. Where plates move apart, ridges, like the mid-ocean ridges, or rifts, like the African rift valley, are apparent. Where plates move past each other, they form transform faults like the San Andreas fault. Where plates move together, one of two things happens. Either ocean crust moves below continental or ocean crust in what are known as subduction zones, or continental crust runs into continental crust, crumpling at the edges and causing collision mountain belts to arise.

continental plate

oceanic plate

plate boundary

tectonic plates
The world is divided into rigid slabs, or plates, that comprise both oceanic and continental crust. They are constantly moving relative to one another, causing changes to the Earth's structure where they meet.

moving jigsaw

As this theory was elaborated, many of the oddities that Wegener had identified early in the 20th century became essential pieces of the jigsaw – for example, the matching shape of coastlines and matching rock strata in continents such as Africa and South America, now separated by thousands of kilometres of ocean. Proponents of the

theory suggested that some continents were disposed to
split apart. As a continent did so, a new ocean would form
in the space created. The expanding ocean floor would
then separate the two new continents, carrying them
apart on its conveyor belt of new rock spreading from
the central ocean ridge (see p.25).

scale of movements

Some of the first ideas about the scale and speed of the
movements of tectonic plates came from estimating the
age of a particular part of the ocean crust and the distance
it had travelled from the ocean ridge, which gave average
speeds for the plate movements. These have now been
assessed in more detail with more accurate magnetic maps
of the sea floor, and combined with direct and indirect
measurements using space satellites. For example, by
bouncing laser beams transmitted from observatories
on two different continents off reflecting mirrors on
satellites, it has been possible to achieve very exact
measurements of changes in their distance apart over
time. The result is an estimate of the relative velocities
at most of the plate boundaries in the world.

   The movement is so gradual that it is not surprising it
was previously imperceptible. Each side of the Atlantic
Ocean, for example, is moving away from the centre at
about 1cm (0.3in) per year, so the United States has

**varying speeds
of movement**
*The Earth's tectonic
plates, like these
spinning plates,
are all moving at
different speeds
and in different
directions relative
to one another.*

plate boundaries

Most tectonic activity takes place at the boundaries between the plates. Plates may be moving towards each other, moving apart, or sliding past each other. Where two continental plates or a continental and an oceanic plate collide, a mountain range forms. A chain of volcanic islands forms where two oceanic plates meet. Where two oceanic plates move apart, magma rises to the surface to form new ocean crust. Where two plates slide past each other, a fault line occurs.

mid-atlantic ridge

san andreas fault

andean mountain chain

plate boundary

transform faults

When one plate slides past another, it forms a transform fault. The San Andreas fault in California is an example of this. It marks where the Pacific Ocean and a little of westernmost North America is moving north relative to the rest of the continent.

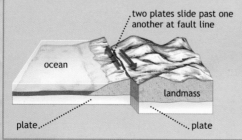

two plates slide past one another at fault line

ocean

landmass

plate

plate

san andreas fault

mid-ocean rifts

When two adjacent ocean plates move apart, a mid-ocean rift valley is formed. Magma rises from the mantle into the rift and solidifies on the surface to form an underwater mountain range, which can reach heights of 3,000m (10,000ft). This breaks into a series of stepped sections as transform faults enable the new sea floor to curve around the Earth. The Mid-Atlantic Ridge rises above sea level in Iceland.

icelandic section of mid-atlantic ridge

oceanic rift valley

direction of ocean plate

mantle

spreading mountain ridge

transform fault

new ocean floor

magma pushes upwards along centre of rift

subduction zones

When an oceanic and a continental plate converge, the denser oceanic plate plunges beneath the continental plate into the mantle. This is known as a subduction zone. The oceanic plate collides with and then moves below the continental plate of South America in this way. As a result of the impact of the collision, the continental crust was crumpled and the Andes mountain range was formed.

andean mountain chain

ocean

subduction zone

oceanic plate plunges below continental plate

mountains thrust up by collision

continental plate collides with oceanic plate

moved nearly 5m (16.5ft) away from Great Britain since the Declaration of Independence in 1776, and gets 2cm (0.75in) further away every year. The Atlantic is spreading relatively slowly. The East Pacific Rise is currently the fastest mover, with a spreading rate of 10cm (4in) per year. And the eastern Pacific Ocean is moving north relative to most of California at around 6cm (2.3in) per year; the boundary through the continent is the San Andreas fault system, and the inexorable movement is responsible for the many earthquakes that occur within this transform fault.

plate to mantle

These numbers, however, show only how the plates move relative to each other. They do not tell us how plates are moving relative to the Earth's core and mantle. An indication of this can be found by looking at sea-floor volcanoes such as Hawaii and its island chain (see also p.59). This line of volcanoes was formed by the movement of the plates dragging the ocean crust over a single hot spot – a plume of hotter rock flowing upwards from deep inside the mantle. Radiometric dating shows that the oldest volcanoes are at the north of the chain, with the younger active volcanoes to the south. The distance between two volcanoes can be taken as indicating the distance the plate has moved relative to the depths of the Earth between the dates when each was formed.

In the early 1960s, the Canadian geophysicist **John Tuzo Wilson** (1908–93) put forward the idea that volcanic ocean islands, such as Hawaii, formed over hot spots in the mantle beneath; chains of such islands reflected the direction of movement of the plates over the mantle below.

Combining such computations with our knowledge of the movement of plates relative to one another gives further insight into this complex process. It shows, for example, that the western Pacific plate is moving fastest

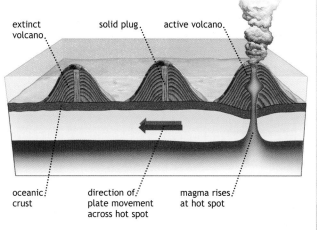

extinct volcano.

solid plug.

active volcano.

oceanic crust

direction of plate movement across hot spot

magma rises at hot spot

volcanic island chain
The Hawaiian island chain is a visible indication of the speed and direction of movement of the plate on which it sits. Each volcanic island was formed as the ocean crust passed over a plume of magma rising at a hot spot in the Pacific plate.

relative to the mantle below, zooming west-northwest at around 10cm (4in) per year, while Eurasia and Antarctica by contrast are scarcely moving at all.

plate geometry

The geometry of the plates and their movements appeared very simple at first sight. Early ideas on the workings of the plates could be – and some of them were – worked out using a piece of paper and scissors. If you cut a piece of paper in two, using a zig-zag line made up of straight segments at right angles, then slide the paper apart and together again, keeping it parallel, you have made your own pair of plates, with ridges, subduction zones, and transform faults. Move it at right angles, and you have another configuration, and you can see how the pattern of ridge, transform fault, and subduction zone link the possible plate configurations and their relative velocities.

However, on the real Earth, plate boundaries are curved, not straight, and plates rotate as they move. So plate tectonic modellers soon began to work with similarly simple models for the plate movements, but using the geometry of a sphere, in which great circles take the place

mountain building

Great mountain chains are formed where tectonic plates collide. The Alps and Himalayas are examples of mountain systems created by a collision between two continental plates. Because they are equally buoyant, both plates crumple, instead of one subducting as happens when a continental plate and an oceanic plate collide.

forming collision mountains

Formed by a collision between the European and African plates, the Alps are "fold" and "fault" mountains. In contrast to the Andes, formed where continental and oceanic plates meet, the Alps are not volcanic.

high mountains built up by repeated folding and faulting

folds become progressively more deformed

foothills where rock layers begin to fold and slip as plates collide

continental plate made of horizontal layers of rock

rifting

Mountainous terrain can also result from rifting. In the Sinai desert, for example, the crust is stretching as the Red Sea opens up. Blocks of rock slip downwards to make long narrow valleys, bounded by steep faults.

a horst is a rock thrust up between faults

a block of rock slipped down between faults is called a graben

steep faults form as the crust stretches

in rift mountains layers of rock remain horizontal

of straight lines. This method made the plate tectonic map much clearer. It also clarified the role of a strange set of fracture zones discovered as the ocean floors were mapped in detail. These fractures cut across the ridges and displace them. On a sphere, their role becomes clear: they allow the ridges to curve around the Earth.

Using geometrical models and the available evidence of past movements, such as the bar-code magnetic stripes of the ocean floor (see p.25), it is possible to track the plates back through time, and to predict what will happen in the future. Pull back the conveyor belt for 60 million years or so and the Atlantic Ocean, for example, closes, and Europe and North America join.

the formation of mountains

Unravelling the plate movements made sense of earlier geological observations in the great fold mountain belts, such as the Alps and Himalayas. When continents meet, neither subsides gracefully into the mantle, because they are each as buoyant as the other. Instead, they collide, and both are distorted by the inexorable movements of the plates. They crumple into huge folds like thick blankets, and fracture along faults that can leave layers of rock stacked up like a pile of books.

The Himalayas, for example, exist because the Indian subcontinent, originally a separate continent, has been moving northwards into Asia for the past 50 million years. The original collision produced spectacular folds and long, almost flat faults called thrusts that pile layers of sedimentary rocks on top of each other. The rocks of the two continental crusts were stacked on top of each other and the crust became much thicker.

India has moved about 1,000km (620 miles) north into Asia and is still moving north today at around 3cm (1.25in) per year. This is why the Himalayan range is still getting wider, as folding and faulting continues to spread

recap

The **ocean conveyor belt** starts at the mid-ocean ridge, where molten rock rises from the mantle and solidifies. The new ocean floor slides away from the ridge to the edge of the ocean, where it bends and slides down into the mantle again.

himalayas
The highest mountains in the word are part of the Himalayan chain, thrust upwards on great folds and faults because India is pushing northwards into Asia and both continents are crumpling.

around the main mountain belt. It is also getting higher. Nanga Parbat, one of the Himalayan peaks, is rising at about 2cm (0.75in) per year, outstripping even the intense erosion found at these altitudes.

making sense of the past

Painstaking detective work, collecting and collating geological evidence from ancient mountain chains, has revealed many long-vanished oceans. The Alps of Western Europe, for example, not only have sedimentary rocks containing the fossils of sea creatures among some of their peaks, but, among the rifts and volcanic rocks now crumpled into immense folds, scientists have also traced the formation of the vanished Tethys Ocean, in which those creatures once lived. There are also signs that the oceans we know today did not always exist. The ancient mountain chains of the Appalachians in the United States and the Caledonian mountains of Scotland match. They are composed of the same rocks of the same ages, which were crumpled up in the same way 450 million years ago. The fossils they contain and the deformation that they have undergone are identical. This suggests they were formed by an ancient continental collision, before the Atlantic formed. There are the remains of volcanoes of the type found in subduction zones (see p.55) along the line of collision, and traces of a long-vanished ocean.

Bringing together observations of all types – including, for example, the fossil record, trapped magnetism, and absolute dating techniques – has allowed maps of the world as it has changed over the past few hundred million years to be established. Although this represents only a fraction of the 3,600-million-year history of the Earth's

> **" There rolls the deep where grew the tree. / Oh earth, what changes hast thou seen! / ... They melt like mist, the solid lands, / Like clouds they shape themselves and go. "**
>
> Alfred Tennyson, *In Memoriam*, 1850

key points

• The Earth's outer shell is made from a few rigid plates, which are constantly moving relative to one another.

• Ocean floor forms where plates move apart.

• Plates that move together make mountains, either from volcanoes or by continental collision.

absolute dating

It is very hard to determine exactly how old rocks are. The main absolute (as opposed to relative) dating techniques depend on the study of radioactive isotopes trapped in minerals in metamorphic and igneous rocks (see p.17). These isotopes begin to decay as soon as they form, emitting nuclear particles and so turning into new isotopes. Because this decay takes place at a fixed rate, comparing the amount of original and new isotope present today gives the time since the mineral – and therefore the rock – formed.

electrons orbiting around nucleus

protons and neutrons in nucleus

isotope
Isotopes are atoms of the same chemical element that have different numbers of particles, such as neutrons and protons, in their nucleus. Some isotopes are unstable and lose particles, becoming new, lighter isotopes.

zircon crystal
Zircon is one of the most useful minerals for dating because it preserves traces of radioactive decay. At 4,300 million years old, zircon is also the oldest mineral on Earth.

sandstone
igneous rock
limestone
shale

igneous time marker
Once a layer of igneous rock has been dated, scientists can estimate the age of rock strata above and below it by relative dating techniques (see p.8).

crust, the maps cover much of the period in which complex life has existed. Wegener's idea that there was once a supercontinent called Pangaea (see p.11) has been explored more thoroughly and vindicated. These maps of the past confirm that the processes that shape the world today have gone on in the past, albeit with different plate boundaries and, indeed, different continents.

recent discoveries

Our understanding of the structure and complex workings of the Earth's crust and mantle is improving all the time, as more information is gathered with ever more sophisticated equipment. In recent years, much exploration has focused on the deep ocean. *Alvin* and the few other similar deep-sea submersibles have visited the sea floor repeatedly. The scientists on board have made some amazing discoveries during these trips. The big surprise came in 1977 on a dive in the Pacific Ocean, near the Galapagos Islands. Researchers were looking for signs of hot springs on the sea floor, reasoning that sea water would seep into the hot volcanic rocks and emerge much hotter, carrying minerals that might produce mineral deposits on the sea floor. They found just these types of mineral deposits, and, surrounding them, they found a cluster of new types of animals, thriving in isolation on the deep ocean floor. There were clams, shrimp, crabs, starfish, and immense tube worms, all surviving in complete darkness on the ocean floor: the only life on Earth that does not depend on the Sun.

diving to the sea floor
Three people can journey to the bottom of the sea in Alvin, the US submersible, squeezed inside a 2m- (6ft-) wide ball that can withstand the immense pressures of the deep-sea floor.

black smokers

On the dark, empty deep-sea floor, there are vents of boiling hot sea water laced with metal sulphides. This is sea water that has seeped into the fractures of the newly formed sea floor and circulated through the hot rock there, dissolving minerals and becoming hotter as it passes through. When the boiling water solution bursts out and makes contact with the cold sea water, the sulphides come out of the solution as black particles, which build up into a chimney-like structure known as a black smoker.

clouds of minerals
The hot plumes of black sea water spurting from the deep-sea vents reach temperatures of around 300°C (572°F). The dark clouds are composed of tiny grains of metal sulphides derived from the newly formed rocks of the mid-ocean ridge.

shrimp

clam

giant tube worm

living in darkness
Blind shrimp, clams, and giant tube worms up to 1m (3ft) long cluster around the black smokers in the Pacific. They depend on specially adapted, sulphur-loving bacteria that live around the black smokers.

hot springs

Hot water full of minerals or sulphur sometimes emerges on land in the form of a hot spring or – if the water is very hot – a geyser. The hot spring below is one of many in Yellowstone National Park. A century ago, mineral-rich hot springs were used as health spas. Microbes known as "extremophiles" thrive in such environments.

Organisms capable of living under extreme conditions have now been found in many underwater vents, around hot springs on land, in deep mines and boreholes, and even in the exposed, dehydrated dry valleys of Antarctica. In fact, some sort of hydrothermal vent is now considered a likely place for life to have started on Earth.

refining the theory

Satellite exploration has resulted in some modification of the simple plate tectonic theory. Satellite images reveal that the oceanic and continental plates do not behave as the perfectly rigid lithosphere bodies imagined by the first theorists. Instead of sharp, narrow boundaries, some parts of the Earth's surface show plate boundaries where deformation spreads over hundreds of kilometres. The effects of the collision of India with Asia, for example, have spread over thousands of kilometres beyond the actual zone of the collision. The whole of this extensive area is marked by shallow earthquakes.

drilling in antarctica
By drilling boreholes, scientists have discovered crucial information about the nature of the rocks in the Earth's crust.

The increase in information about the Earth has also consolidated what researchers already knew about the driving forces of plate tectonics. Many of these advances owe a debt to better computing power, both for data collection and for modelling and data analysis. Modern techniques, such as seismic tomography, in which powerful processing methods use seismic waves from individual earthquakes to build up a three-dimensional picture of the Earth's mantle, confirm that the descending slabs of oceanic crust in subduction zones (see p.33) remain intact through the upper mantle.

looking at other planets

Researchers have also been examining other planets to see if plate tectonics is a feature of all rocky planets. It seems that it is not: the surfaces of Mars and Venus are dominated by volcanoes but do not show signs of the distinctive movement patterns of the tectonic plates on Earth. However, there may be analogous processes in the outer reaches of the solar system. Some of the giant gas planets, such as Jupiter, have moons that are made of frozen water, like ice on Earth, together with frozen ammonia, sulphur compounds, and other mixtures. Some of these moons, such as Europa, show rafts of solid ice edged by bands of ridges, and rifts, like continents edged by rifts and collision belts. Perhaps "extremophiles" (see p.41) might live here.

key points

• Black smokers, supporting specially adapted creatures, proliferate on the mid-ocean ridges.
• Modern techniques have refined plate tectonic theory.
• As yet, plate tectonics has not been found to be a feature of any other rocky planet.

extraterrestrial volcanoes
Mars is marked by huge volcanoes that shaped its surface. Although they have a lot in common with Earth volcanoes, they are not distributed in such a pattern that would suggest that Mars also has tectonic plates.

danger zones

The slow but inexorable movement of the rigid continental and oceanic tectonic plates has transformed the surface of the Earth over the past 4.6 billion years, creating oceans, mountain ranges, and island chains. However, their movement has been most dramatically felt by humankind in the form of earthquakes and volcanic eruptions. Such alarming and devastating activity occurs overwhelmingly along plate boundaries. The theory of plate tectonics, together with advances in modern technology, has enabled scientists to determine the location and probable intensity of earthquakes around the globe. They can also predict with considerable accuracy when and how a particular volcano will erupt, so that those living in the danger zone can take practical steps to deal with the hazard. Long-term planning is possible now too: houses can be designed and built to withstand the impact of most earthquakes. Scientific advances will continue to be made to iron out any anomalies, and to make living on this restless Earth a less dangerous proposition.

volcanic eruption

Jets of magma erupt from this fire fountain volcano in Hawaii. Ash and pieces of the rocky cone are blown into the air at the same time. The runny lava can reach up to 200m (660ft) high, and spatters to the surrounding ground. Some drops of molten rock solidify in the air to land as volcanic bombs.

earthquake tremors

The movement of tectonic plates is generally felt in the form of earthquakes. In 1961, geophysicists across the world, led by US scientists, set up the Worldwide Standardized Seismometer Network to determine the exact location of earthquakes. The same type of seismometer was set up at locations all around the world, all wired to the same clock, and all responding in exactly the same way to earthquake vibrations.

As the earthquakes were plotted on the world map, it became clear that vast areas of continent and ocean floor were untroubled by earthquakes, but that these areas were bounded by zones where the epicentres of earthquakes, occurring over just a few decades, clustered like flies. It also became apparent that many, although not all, of the places prone to earthquakes also harboured volcanoes.

earthquake map
The red circles on this map represent earthquake zones. Almost all quakes happen at the edges of tectonic plates, where the plate movement takes place on a network of faults that cut through the rocks.

fracturing shallow rocks

Careful and precise seismology revealed that earthquakes happen overwhelmingly along plate boundaries. The steady movement measured by averaging out the growth of the ocean floor, or assessed from satellite surveying, masks the fact that, in the crust, the displacement takes place in jerks of individual movement on faults. It is the movement of these faults that causes earthquakes. Most earthquakes originate in the top 20km (12.3 miles) of the

Earth's crust, where it is cold and rigid. The stress that is generated in rocks as the plates slide past each other in transform faults, or as one plate slides beneath another in subduction zones, creates fractures that cause earthquakes.

At the mid-ocean ridges, the newly formed crust fractures to allow more molten rock to reach the surface; it cracks as it cools and also as it moves away from the valley in the centre of the ridge. All these fractures produce almost continuous earthquakes, which mark out the mid-ocean ridges clearly by their seismology.

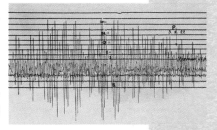

Transform faults also have quakes, arising in the main from fault movements that are roughly horizontal. The plate boundary that cuts through California is a transform fault, and the San Andreas fault system has shaped the landscape through which it runs (see p.32).

Subduction zones too have shallow quakes in the overlying rocks. Some of these represent compression, as the ocean plate moves towards the continent. The continental rocks can be folded and faulted as a result, and the sediments on top of the ocean-floor basalt are scraped off and piled up in a great wedge of sediment against the edge of the overlying continental plate on one side of

tokyo quake
The jagged movements of the needle on this seismograph show the vibrations of the ground (seismic waves) recorded far away from the origin of the quake that occurred on 1 September, 1923.

tsunami
Undersea earthquakes set up waves that travel thousands of kilometres, becoming dangerous when they approach the shore; Pacific islands, such as Japan and Hawaii, are especially vulnerable.

earthquakes

Earthquakes occur when rocks fracture underground as two tectonic plates judder past each other in a transform fault (see p.32) that has become slightly stuck. The jolting sets up vibrations or shockwaves underground that travel through the Earth, and up to and along the surface. The longer the rupture, the bigger the quake. Big earthquakes, such as the 1906 San Francisco earthquake, involve a few metres of displacement along many kilometres of fault.

direction of plate movement

as shockwaves travel away from focus, they cause less damage

epicentre

concentric circles of shockwaves radiate outwards and upwards from focus

focus: starting place below ground

mercalli scale

This practical measure of the size of an earthquake is based on the effects apparent at the surface; for example, ground shaking, collapsing buildings. Californian earthquakes usually involve horizontal slip along one of the many faults that run parallel to the coastline.

focus and epicentre

An earthquake starts where the underground rupture begins, at the focus or hypocentre. The epicentre is the spot on the surface directly above the focus.

XII	total destruction: waves on surface
XI	railway tracks bend; roads break up
X	buildings destroyed; large landslides
IX	general panic; foundations damaged
VIII	chimneys fall; cracks in wet ground
VII	difficult to stand; plaster, tiles fall
VI	difficult to walk; windows break
V	liquids spill over; sleepers awake
IV	dishes rattle; trees shake; cars rock
III	hanging objects swing
II	people at rest notice shaking
I	vibrations recorded by instruments

california
nevada
san francisco
epicentre

the deep ocean trench. These sea-floor sediments pile up on faults, giving rise to many earthquakes, dangerous because they are often close to land. Many of the frequent earthquakes that plague Japan originate in such a wedge.

deep earthquakes

Seismologists discovered to their surprise that, among the earthquakes that occur in shallow rock at the edges of the oceans, there was a distinct group of quakes that formed deep in the crust, hundreds of kilometres below ground. At this depth, the greater pressure and temperature should mean that the rock is flowing, and that any stress caused by plate movements should merely stretch and squeeze it. Pliable rock does not fracture, so what could be causing these deep earthquakes? The seismologists noticed that the deep earthquakes lay in broad bands next to the lines of volcanoes that mark the edge of the Pacific, and other oceans. Precise location of the foci of these earthquakes showed that the quakes were shallowest the nearer they were to the ocean, becoming gradually deeper the further inland they occurred.

It was **Charles F. Richter** (1900–85) who first devised a way of comparing the sizes of earthquakes in different parts of the world. In the 1930s and 1940s, he devised the Richter scale. The scale is logarithmic: a quake at 5 on the Richter scale releases 30 times as much energy as one at 4.

It transpired that these deep earthquakes marked the position of the slab of ocean crust as it slid beneath the continental crust. The earthquakes exist at

the mercalli approach

Giuseppe Mercalli (1815–1914) was a professor of natural sciences who based his earthquake intensity scale (see panel, left) on his own observations in regions of Italy plagued by quakes.

cold caramel
Like lava, caramel shatters when it is cold, but when warm it is a soft, stretchy substance that flows.

unusual quake
The biggest earthquakes in the mainland United States took place in 1812 around the town of New Madrid, Mississippi, far from any known active faults.

such depths because they do not originate in the flowing mantle rocks usually found there: the slab of ocean crust is the origin. Even though it is within the much hotter mantle, the cold, dense ocean crust takes a long time – perhaps 60 million years – to warm up and is consequently still solid when it has to slide under the continental crust. As it cannot bend, it moves by cracking at intervals; it is these fractures that create the earthquakes.

new puzzles

Researchers have recently become aware of a handful of earthquakes that happened within plates, far from plate

boundaries and, it appears, far from any significant sources of stress. One of the most enigmatic examples is the earthquakes that occurred in New Madrid, Mississippi, in 1811–12. These three quakes are the largest ever recorded in the United States, apart from those in Alaska. They are estimated to have had magnitudes of around 8 on the Richter scale (see p.49), and would be major disasters if they happened today. No one knows why these earthquakes happened, nor whether or when they will happen again. Scientists assume that there must be some sort of weakness in the lithosphere to trigger such quakes, but their mechanism is still to be discovered. Surprisingly, despite their magnitude, the earthquakes probably killed only around 100 people.

key points
• Earthquakes happen when rock shatters and slips along a fault zone.
• Most earthquakes happen around plate boundaries.

volcanic activity

At a volcano, magma (hot molten rock, created by the partial melting of the crust and mantle) is able to rise through the crust to the Earth's surface where it is ejected as lava. There are around 1,300 active volcanoes on land, as well as many volcanoes under the sea along the mid-ocean ridges, where eruptions are almost constant. The theory of plate tectonics has made sense of the distribution of volcanoes around the world. Active volcanoes mostly lie on the boundaries between plates, notably in the "Ring of Fire" around the Pacific Ocean and in a belt from the Mediterranean to Indonesia. The nature of the plate boundary dictates the type of volcano, and explains why some lava flows are faster than others and some volcanoes liable to erupt with explosive violence.

what causes eruptions

The most volcanically active places on Earth are the mid-ocean ridges. Here the lithosphere is unusually thin and the mantle unusually high. It is therefore under less pressure than normal and is able to melt at a temperature that would leave it solid at greater depths and pressures.

pacific basin

ring of fire
Spectacular eruptions make a non-stop geological show at the ring of volcanoes around the Pacific Ocean.

The melted mantle, or magma, bubbles up in a mild outpouring of syrupy lava that forms basalt ocean floor as it cools. By contrast, the magma that emerges at subduction zones erupts explosively from chambers deep underground. Huge quantities of carbon dioxide gas is dissolved in the magma under intense pressures underground. As the mixture rises to the surface and the pressure falls, bubbles form in the magma, gradually becoming larger and larger, and finally forcing the magma out in a dramatic eruption. The hot-spot volcano is the only type that does not occur at a plate boundary. Such volcanoes may simmer for a long time above particularly hot spots in the mantle, steadily oozing lava and hot gas.

The particular combination of pressure and temperature existing at the point where an eruption occurs affects the chemical composition of the magma, and therefore the nature of the lava and ash an eruption produces, and the violence or otherwise of the eruption. The speed at which lava flows depends on factors such as its temperature, the pressures built up in the volcano, and the composition of the lava itself – its viscosity, or stickiness. The more silicon and oxygen there is in the rock, the more viscous the lava will be and, consequently, the slower it will tend to flow. Plate tectonics, combined with knowledge of the chemistry of the rocks, can indicate where to expect unusually speedy lava.

volcanoes in the ocean

Direct study of underwater volcanic activity has until recently been impossible, but Iceland, where the Mid-Atlantic Ridge rises above the sea surface, has long shown something of how the ridges work. On this island, made almost exclusively of

cool minerals from hot magma

Hot magma rises from underground chambers, like waxy blobs in a lava lamp, towards a vent in the surface. When the magma emerges above ground, it cools to form igneous rock, and the minerals within it crystallize. Valuable sulphides, such as galena (lead ore), can be found in such rock.

river of lava
Lava flows at several kilometres per hour while it remains hot. The lava cools first at the edges, helping to keep the lava in the centre hot and flowing. Once the river spreads out, all the lava cools and slows down.

basalt, volcanoes are the rule, not the exception, erupting from craters and from long fissures running parallel to the central line of the ridge. Steep faults also run parallel to the ridge, breaking up the rock into blocks that gradually spread apart, widening the island in the same way that the ocean floor spreads.

In more recent years scientists have been able to observe the same processes taking place under the oceans. Small, tough submarines that can dive to a depth of 5km (3 miles) have now thoroughly explored the mountain ranges of the mid-ocean ridges (see p.40). The ridge mountains rise about 2.5km (1.5 miles) above the deep ocean floor and their crests another 2.5km (1.5 miles) higher, on average. Along the crests of the mountain ridges run rift valleys as deep as the Grand Canyon. It is in these valleys that most of the volcanoes that build the ocean floor erupt, the magma welling up from the mantle into the rift left as the ocean floor pulls apart. The magma solidifies almost as soon as it touches the cold water, forming basalt, a tough black rock.

The scale of volcanic activity taking place along the mid-ocean ridges is extraordinary. The basalt from the ridge volcanoes forms two-thirds of the Earth's surface.

ocean volcano
Undersea volcanoes interrupt the rocky pattern of sea-floor mountain ridges with characteristic cones and craters, as shown in this sonar image.

In the 1930s, American geophysicist **Maurice Ewing** (1906–74) pioneered marine seismic techniques. He ascertained the global extent of mid-ocean ridges and discovered the deep central rift in the Mid-Atlantic Ridge in 1957. His surveys of the ocean floors proved that lava erupting from a central rift spreads out to form new ocean floor.

The amount of ocean floor formed each year by ocean spreading amounts to around 4 cu. km (141 billion cu. ft) of lava, erupted along the 60,000km (37,000 miles) of ocean ridge. That amounts to 70 cu. m (2,472 cu. ft) of lava for every metre of ridge that is formed.

rift-valley volcanoes

Volcanoes that occur where continental crust rather than oceanic crust is pulling apart can produce unusually fast-flowing lava. The Great Rift Valley in east Africa is an example. This complicated geological structure is splitting to make new oceans in places such as the Gulf of Aden. Elsewhere, such as in Rwanda and the Dem. Rep. of the Congo, there has been rifting without actual splitting for millions of years.

Volcanoes in rift valleys do not erupt with great violence, but produce lava flows that are especially hot and runny. This is because there is relatively little silica and oxygen in the rocks here to make the lava more viscous. Consequently, lava of this type, at a temperature of about 1,000°C (1,832°F), moves through the landscape at around 60kph (37mph), knocking down or flowing over obstacles in its path. This is extremely dangerous – molten lava is just

lava flood
In January 2002, lava from the eruption of the Nyiragongo volcano flooded into the Congolese town of Goma, destroying many buildings there and filling others with lava.

as dense as solid rock. Volcanoes that occur at other types of plate boundary produce basalt lava that flows at 10kph (6mph) or less. Such a rate of movement is slow enough for people to evacuate houses in the path of the lava.

subduction-zone volcanoes

Subduction zones are also marked by volcanoes, but these erupt in very different ways from rift volcanoes. Subduction-zone volcanoes, such as those in the Andes, form because the mantle beneath them has melted as a result of subduction. The basalt slab of ocean floor is cold when it starts to subduct, but heats up slowly as it slides down into the mantle. First the sediments that collected on the top of the basalt slab – the mud and sand of the sea floor – warm up. These sediments contain water and minerals such as calcium carbonate (limestone). When the sediments heat up, they dry up, releasing water and other volatile components, such as carbon dioxide, into the mantle. They change the composition of the mantle, so that it starts to melt at temperatures and pressures in which normal mantle would be stable as a solid. The subduction-zone volcanoes result from this melting.

The molten magma percolates up through the mantle and accumulates in the crust. But, because the crust and the lithosphere of the continents is so much thicker than that of the oceans, it takes longer for the magma to travel upwards at subduction zones than at mid-ocean ridges. The magma also spends longer within the thicker continental crust. Both these effects can alter the chemistry of the magma, producing rock with much

recap

When an oceanic plate meets a continental plate, the denser oceanic plate dives below the continental plate, forming a **subduction zone**.

andes volcano
Subduction of the Pacific plate beneath South America has built the Andes mountain chain from many generations of volcanoes, and is responsible for the numerous active volcanoes there today.

erupting volcanoes

Most volcanoes happen on plate
boundaries. They occur in rift valleys in
mid-ocean ridges and between two
continental plates; here the mantle wells
up and melts, oozing out at the surface.
They are found where an ocean plate
subducts beneath a continental plate or
another ocean plate; here the chemistry
of the mantle is changed so it melts and
the intense pressure causes an explosive
eruption. A few volcanoes occur in the
ocean floor well away from any plate
boundary, where a hot spot in the mantle
sends up plumes of magma that burn
through the lithosphere like a blowtorch.

eruption at mount etna
*Magma explodes from the crater
in a spectacular fountain of fire
in this subduction-zone volcano.*

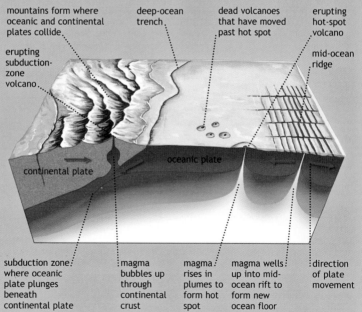

mountains form where
oceanic and continental
plates collide.

deep-ocean
trench.

dead volcanoes
that have moved
past hot spot

erupting
hot-spot
volcano

erupting
subduction-
zone
volcano.

mid-ocean
ridge

continental plate

oceanic plate

subduction zone:
where oceanic
plate plunges
beneath
continental plate

magma
bubbles up
through
continental
crust

magma
rises in
plumes to
form hot
spot

magma wells
up into mid-
ocean rift to
form new
ocean floor

direction
of plate
movement

subduction volcano

Magma forms deep in the Earth. It seeps slowly upwards and collects in a chamber below a volcano. From there it rises towards the surface, etching out a pipe. As it rises, the pressure drops enough for the gas dissolved in it to escape, like bubbles in a fizzy drink. The bubbles increase the pressure, which finally blows the magma apart, and the volcano erupts.

magma flows along branch pipe

main volcanic pipe

layers of lava and ash build up during eruption

hot springs occur close to volcano

magma forced up main pipe by pressure below

magma collects in chamber underground

types of volcano

The shape of a volcano depends on the way it erupts, which in turn depends on how it formed. A volcano erupting thick, sticky lava will produce a dome. If the lava can flow more easily away from the crater, a shield volcano forms. Fissure eruptions result when the magma exploits a line of weakness in the crust.

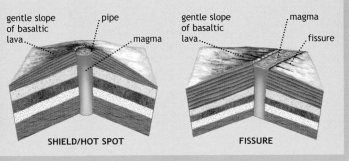

steep convex slope from thick, fast-cooling lava

pipe

magma

DOME

gentle slope of basaltic lava

pipe

magma

SHIELD/HOT SPOT

gentle slope of basaltic lava

magma

fissure

FISSURE

more silica and oxygen than the ocean basalts. These magmas are much more viscous – some of them scarcely flow, once erupted. Instead they form lava domes.

The viscosity of the magmas is responsible for the violent behaviour characteristic of subduction-zone volcanoes. Because they are so much more viscous, they can support far greater pressures before erupting. But when they do erupt, they explode. These immense eruptions blow molten and solidified rock into tiny fragments, which fall as ash, or build plumes that can reach into the stratosphere and spread right around the world.

> **"The whole north side of the summit crater began to move instantaneously as one gigantic mass."**
>
> Dorothy Stoffel, Mount St. Helens eyewitness, 1980

Mount St. Helens in Washington State, USA, is a subduction-zone volcano of this type. It exploded in May 1980 when the rising magma within the volcano bulged out the sides of the mountain so much that a landslide began near the summit. As the rock sheared away, the pressure on the molten rock within dropped suddenly and it exploded, blasting out across the mountainside in a thundering avalanche of hot ash and gas that travelled 20km (12 miles) from the crater. Many of the volcanoes around the Pacific Ocean are subduction-zone types, and produce the thick, sticky magmas that tend to explode.

mount st. helens
The 1980 eruption of this volcano started with a landslide at the edge of the crater; a hot roiling cloud of ash and debris then blasted down the mountainside, flattening trees and filling lakes. Ash fell across the US.

island-arc volcanoes

Some of the Pacific island chains – for example, Tonga and the Aleutians – involve subduction of a slightly different type to the Andes. They are in regions where one ocean plate descends beneath another ocean plate. It is difficult for scientists to understand how such a system starts, without the difference in density that sends the denser ocean crust beneath the buoyant continental crust. However, once established, such oceanic subduction zones can result in a chain of volcanic ocean islands which, if they erupt, risk triggering tsunamis that spread devastation over a vast area.

hot-spot volcanoes

There is also a type of ocean island volcano that does not occur at plate boundaries. These are known as hot-spot volcanoes. Sea-floor volcanoes, such as Hawaii and its chain of islands, for example, are situated well away from the boundary of their oceanic plate. Researchers think that the mantle beneath them is unusually hot because a plume of hotter rock from deep inside the mantle – maybe even from the edge of the Earth's core – for some reason flows upwards. This is called a mantle plume.

The process of creating a chain of hot-spot volcanic islands has been likened to moving a piece of paper over a candle flame – as you pull the paper along, you get a line of scorched spots. Similarly, as the movement of the tectonic plate drags the ocean lithosphere over the immobile hot spot, every now and then the hot spot burns through the lithosphere, making a volcanic island (see p.35). The oldest, inactive volcanoes in the chain are always furthest from the hot spot, and the young, active ones nearest to it.

key points

• Volcanoes form at plate edges and above hot spots.
• The plate tectonic pattern shows what type of volcano to expect where.
• The effects of a volcanic eruption can spread around the world.

hot spot
Volcanic islands such as Hawaii form where a plate moves over a hot spot in the mantle beneath. The hot mantle makes a volcano, just as a candle scorches a sheet of paper held above it.

living with danger

However life started on Earth, it has survived, and has done so on an ever-changing surface. People have often been prepared to live close to sources of danger, such as volcanoes and faults that cause earthquakes, because of other benefits, such as soil fertility, that the area offers. Italy has no shortage of volcanoes and there are frequent earthquakes that act as warning signs of a forthcoming eruption, yet the Roman Empire began there and people have thrived there ever since.

impact of volcanic eruptions

Even relatively small eruptions, such as the 1980 eruption of Mount St. Helens, cause immense damage. The really enormous eruptions of the past are difficult to imagine. The most recent eruption of the huge caldera in Yellowstone National Park, some 600,000 years ago, covered the United States with ash from California to the Gulf of Mexico, and up to the Canadian border. It produced about 1,000 cu. km (35 billion cu. ft) of ash and rock: 40 times as much as the 1980 eruption of Mount St. Helens (see p.58).

Eruptions can cause changes to the weather. One of the biggest recorded eruptions took place at Tambora in Indonesia, in 1815. The dust and gases released spread through the atmosphere, reducing the amount of the Sun's energy that reached the Earth. It was held responsible for two years of poor weather around the world, making 1816 the "year without a summer" for large parts of eastern North America, western Europe, and China. Tree ring evidence supports the climatic impact of the Tambora volcano, revealing that these were two poor growing years. In 1991, the eruption of Mount Pinatubo in the Philippines also reduced temperatures on Earth for about two years.

Earth movements, in turn, can affect the climate. The growth of the Himalayas and the development of the Tibetan plateau was instrumental in establishing today's monsoon in the Indian subcontinent. The plateau is an important part of the monsoon system now. As the temperature rises in summer, atmospheric pressure drops, and warm, moist air flows in from over the oceans, bringing the rains. The Tibetan plateau is a young feature, geologically speaking. It is now about 5,000m (16,400ft) above sea level and probably gained most of this height between about 25 and 15 million years ago. As the plateau rose, it got in the way of the high-altitude air circulation, changing the path of the jet stream. In so doing, it changed the world climate system.

predicting quakes

The study of plate tectonics continues to clarify the pattern and nature of earthquakes around the world. It is now known that at a transform fault, such as the San Andreas fault in California, the fracturing that creates an earthquake and allows the plates to move past each other, does not account for all the plates' movement. Some of it is taken up in tiny steps and infinitesimal distortion called creep. Researchers can identify places where the

icy dino
At the end of the Cretaceous Period, dinosaurs were mysteriously wiped out. One theory is that a mass of volcanic eruptions occurred, perhaps due to a meteorite hitting the Earth. As a result the Earth cooled dramatically and plants were coated with dust and ash. Lack of sunlight killed the plants, starving the dinosaurs as they shivered and choked in the cold, polluted air.

combination of quakes and creep are not keeping pace with the plates: these are the places to expect earthquakes. What does not seem possible, at present, is to say with any certainty when a quake will come, and there are no irrefutable signs that an expected quake is imminent.

forecasting volcanic activity

It is possible to tell when a volcano is getting ready to erupt because of the physical processes that must precede an eruption, such as bulging and the emergence of steam plumes. Most of these can be monitored from a safe distance, so that vulcanologists can assess the behaviour of the volcano in an active phase and evacuate people from the areas around volcanoes before they erupt. But not all volcanoes behave as expected; they still spring surprises.

> **A major volcanic eruption has the capacity to affect a significant fraction of Earth's population.**
>
> Ernest Zebrowski, vulcanologist, 1997

And not all volcanic hazards come directly from eruptions. Pyroclastic flows – huge hot avalanches of dust and gas – can kill people and destroy buildings kilometres away from the slopes of a volcano. Volcanic ash makes very fine mud when mixed with rainwater, making the slopes of volcanoes unstable. Combine heavy rains with frequent earth tremors and you get devastating volcanic mudslides – lahars – that can travel at speeds of 40km (25 miles) per hour.

mudslide
Volcanic ash makes for sticky mud deposits around active volcanoes; add rain and you have the right conditions for mud avalanches, known as lahars, which have caused devastation on the slopes of Mount Unzen in Japan.

predicting an eruption

Warning signs before a volcano erupts can be monitored, often from a safe distance, and used to assess the potential threat. As the magma rises underground, the volcano bulges and sets off small earthquakes. It also becomes hotter, gives off gas and steam, and sometimes lava and ash.

studying a volcano
Special heat-resistant clothing enables vulcanologists to get close enough to lava to take samples for analysis.

steam plumes visible

gas seeps from cracks

sides of volcano buldge

tremors occur

signs of molten rock

water temperature rises

use of hazard maps

In an attempt to predict the potential dangers of living in a volcanic or earthquake zone, geologists prepare hazard maps. They put together all the available information for a particular area to predict the nature and extent of any future volcanic or earthquake activity there. For example, they might look at the rocks formed in ancient volcanic eruptions at the same spot, and identify which valleys would channel lahars or pyroclastic flows, and where they

might come out. The city of Seattle, for example, is partly built on a prehistoric mudflow that swept across from Mount Ranier, suggesting that similar processes in a future eruption will pose a similar threat. For earthquakes, such as those threatening the city of San Francisco, hazard maps take into account the distance from the likely earthquake fault and the nature of the rock and soil immediately below the surface. The vibrations that damage buildings are much worse in loose soils and rubble, compared to solid rock. In addition, the shaking makes sandy layers in the subsurface liquefy so that they behave like quicksand, leading to irregular subsidence and collapsed buildings.

heavy damage
Houses in the Marina district of San Francisco suffered particularly in the 1989 quake because they were built on soft ground.

surviving a strike

Now cities such as San Francisco are much better able to stand up to earthquakes, thanks to the ingenious ways in which engineers can now construct and outfit buildings. Engineers can design buildings that flex in an earthquake, rather than collapse, or can set buildings on rubber dampers to absorb the worst of the shaking. Traditional styles of architecture in countries where earthquakes are common, such as Japanese wooden-framed houses, are frequently the model for the safe buildings of the future.

earthquake-proof
Steel bracings and a coating of liquid concrete have been retro-fitted to this building to make it resistant to quakes.

But there is often very little that anyone can do should an earthquake strike, especially in an area where they are rare. And the measures to anticipate and mitigate the effects of quakes are expensive. Most volcano and earthquake areas are in developing countries, where there may be other priorities than the intensive monitoring or costly building modification that makes a difference. In developing countries, the strategy has to be to focus on saving lives when an quake hits.

practical understanding

Throughout history, earthquakes and volcanoes have been seen as terrifying natural happenings that come out of the blue and strike at random. Modern earth science can offer understanding, some measure of prediction, and practical strategies for easing their effects. But more than this, it can assert with confidence that there is a pattern to the restlessness of the Earth. Plate tectonics has transformed the map of the globe in more ways than could have been imagined 100 years ago. Continental drift and mantle convection have become part of the established views of geologists today, and, through plate tectonics, they make sense of many strands of information about the Earth's surface. They provide a strong foundation in the pursuit of further discoveries and in the service of the people who live on this restless Earth.

❝ It is our good fortune to ⸤li⸥ve in an era that has enabled ⸤m⸥ankind to understand the ⸤E⸥arth far more clearly than ⸤e⸥ver before. ❞

⸤W⸥alter Sullivan, science journalist, 1991

key points

• Understanding how earthquakes and volcanoes form helps scientists to forecast their behaviour.
• Technological advances make living with earthquakes and volcanoes safer.

observing lava
Vulcanologists have gained much of their knowledge about the nature of volcanoes and the lava they emit by watching them in action. Their increasing understanding enables us to avoid the worst effects of volcanic eruptions.

glossary

ash
Dust-sized fragments of rock blown apart in a volcanic eruption. Ash usually falls around the volcano and can build up into a cone. After a big eruption, it can spread around the world.

asthenosphere
The soft centre section of the mantle. The rigid upper section of the mantle and the crust above it (which make up the lithosphere) float on top of this fluid layer.

core
The innermost part of the Earth, made mostly of iron. The inner core is solid; the outer core liquid.

crust
The rigid outermost layer of the Earth, made of different rock and having different physical properties from the mantle below.

crystal
A natural form of a mineral in which the elements are arranged in regular three-dimensional patterns. Crystals have particular shapes and colours that are characteristic of particular minerals.

earthquake
A series of vibrations occurring within the Earth, set off by some sudden movement, typically slip on an underground fault.

earthquake epicentre
The point underground where an earthquake originates.

earthquake focus
The point on the surface of the Earth directly above the spot underground where an earthquake originates.

fault
A discontinuity in the Earth's crust, where the rocks on either side slide past each other.

fossil
The remains of an animal or plant, preserved in rock. Often minerals have taken the place of the organism, or replaced the original material, so that the shape is preserved in detail. Some fossils are trace fossils. These preserve the tracks made by a long-dead animal.

geological timescale
The timeline of the Earth's history, constructed from a combination of relative and absolute dating techniques.

hot spot
An area of hotter than normal rock in the Earth's upper mantle that results in the occurrence of a volcano on the Earth's surface above it. A hot spot is often the result of a mantle plume rising from deeper within the molten part of the mantle.

hydrothermal vents
Places on the sea bed where water heated by circulating through cracks in the rocks of the sea floor escapes as jets laden with minerals. These jets can form plumes of cloudy water known as black smokers.

igneous rock
Rock formed when molten rock from the mantle solidifies on the Earth's surface.

lava
Molten rock flowing over the surface of the Earth that has erupted from a volcano.

lithosphere
The rigid outer layer of the Earth, comprising the upper part of the mantle and the crust above it. It is split into tectonic plates.

magma
Molten rock lying underground, and originating in the mantle, before it erupts in volcanoes.

magnetic mineral
A mineral, such as magnetite, which can be magnetic and act in the same way as a tiny magnet.

magnetic reversal
The term used when the Earth's magnetic field reverses direction. For example, when a compass needle that had pointed towards the geographical north pole reverses to point south. This

has happened many hundreds of times in the history of the Earth.

mantle
The rock layer that lies beneath the crust and above the core of the Earth. It is made of different minerals and is so hot that it is partially molten.

mantle plume
A buoyant body of hot magma that rises from the mantle, creating a hot spot in the crust above it, which triggers a volcanic eruption.

metamorphic rock
A rock in which the original minerals have changed into new configurations as a result of enduring high temperature or pressure after being buried or forced next to molten magma.

mid-ocean ridge
The mountain range found along the spine of each major ocean. Here new ocean crust forms, bubbling up from the mantle, as two oceanic plates move apart.

mineral
An element, or compound of different elements, found in rocks. Each mineral has characteristic chemical composition and crystal structure.

palaeomagnetic
Traces of the ancient magnetic field of the Earth found in rocks that contain magnetic minerals.

radioactivity
The emission of radiation and particles from some elements that are naturally unstable.

radiometric dating
Discovering when a rock crystallized or solidified by measuring the proportions of particular radioactive elements within it.

remote sensing
The measurement of physical properties of the Earth and other planets from a distance.

rift valleys
Great valleys formed when two continental plates start to split apart, sometimes as the first stage in the formation of a new ocean.

seamount
An isolated extinct volcano found underwater; most were once islands, but are now submerged as a result of the ocean floor subsiding as it grows older.

sediment
Rock debris, such as sand or mud, that accumulates on the sea floor or in rivers.

sedimentary rock
Rock formed by the solidification of sediments, such as sand or mud.

seismic wave
Vibrations that travel through the Earth, usually as a result of an earthquake. Mine collapses, explosions, and even rock concerts can also set up seismic waves.

seismology
The study of the Earth using seismic waves, either from earthquakes or from controlled explosions.

seismometer
An instrument to detect seismic waves, to find out more about earthquakes or the structure of the Earth.

stratigraphy
Documenting the order in which layers of rock formed, and developing a geological timescale based on this relative method of dating rocks.

subduction
The process that occurs when one tectonic plate slides under another, generating volcanoes and deep earthquakes in the process.

subduction zone
An area where one tectonic plate slides beneath another. Where an oceanic plate slides beneath a continental plate, volcanoes and mountain ranges result. Where an oceanic plate slides beneath another oceanic plate, lines of ocean islands are created.

tectonic plate
One of about nine large and a dozen smaller, rigid moving sheets that make up the Earth's lithosphere. They may be moving past each other, towards each other, or away from each other.

transform fault
A plate boundary where two plates slide past one another.

volcano
An eruption on the Earth's surface of magma from the mantle, which has risen through a rift in the crust.

index

a

Africa, coastline matching South America 12, 13, 30
Alaska, earthquakes 50
Aleutians 59
Alps, formation 36, 37
Alvin (submersible) 40
ammonites 9
Andes:
  formation 33, 36, 55
  volcanoes 55, 59
animals:
  deep-sea 40-2
  evolution 6
Antarctica:
  drilling in 42
  movement 35
Appalachians 38
ash 52, 58, 62, 66
Asia 5
asthenosphere 30, 66
Atlantic Ocean:
  depths 21
  history 35
  mountain range in 22
  movement 31-4
atolls 24

b

bacteria 41
basalt 15, 16, 25, 52, 53
Bikini atoll 24
black smokers 41, 43
boreholes 42
brachiopods 8, 9
buildings: earthquake-resistant 64-5

c

calcium carbonate 55
Caledonian mountains 38
Cambrian period 8
carbon dioxide 55
Carboniferous period 8
Caribbean 24
Carlsberg Ridge 22
Challenger Deep 22

Challenger, HMS 21
clams 40, 41
climate:
  affected by earth movements 61
  changes in 10
coal seams 10
coastlines, matching 12, 13, 30
Colorado Plateau 16
Congo, Dem. Rep. of 54
continental drift 10-12, 13, 23, 26
continents: rise and fall 16-17
convection, effects of 27
convection cells, in mantle 26, 27
conveyor belt, ocean 25, 37
coral atolls 24
coral reefs 24
corals 8
core 14, 15, 66
  plate movement relative to 34
crabs 40
creep 61-2
Cretaceous period 9, 61
crinoids 9
crust 14, 15, 66
  composition 16
  layers 29
  ocean 50
crystalline rocks 15
crystallization 17
crystals 16, 66

d

Dana expedition 22
darkness, organisms living in 40-2
Darwin, Charles 6
dating techniques 4, 7-10, 19, 23, 67
  absolute 39
Devonian period 8
Didymograptus graptolites 8
dinosaurs 10, 61

e

Earth:
  age 7, 10
  history 8-9
  layers in 14, 15
  origins 7
  structure 15
  temperature 15
earthquakes 4, 13, 14, 28, 44, 46-50, 66
  buildings resistant to 64-5
  causes 34, 48, 50
  deep 49-50
  epicentre 46, 48, 66
  focus 48, 66
  location 46-7
  map of 46
  measuring 48
  on mid-ocean ridges 47
  plotting 46
  predicting 44, 61-2
  shallow 42, 46-7
  as signs of eruptions 60
  surviving 64-5
  undersea 47
  volcanoes linked to 46, 49
  within plates 50
  zones of 46
East Pacific Rise 22, 34
echinods 9
echo sounders 22
erosion 16
Etna, Mount 56
Eurasia, movement 35
Europa 43
evolution 6-7
Ewing, Maurice 54
extremophiles 41, 43

f

fault mountains 36
faults 14, 30, 32, 46, 66
ferns, fossils 10
fire fountain volcano 45, 56
fissure eruptions 57

fold mountains 36
fossils 8–9, 10, 16, 28, 38, 66
  index 9
fracture zones 37

g

galena 52
gastropods 9
geological timescale 8–9, 66
  magnetic 24
geysers 41
Glossopteris fern 10
gneiss 16, 29
Goma, Congo, lava flood 54
Gondwana 11
Grand Canyon 16
granite 15, 16, 29
graptolites 8
Great Rift Valley 54
Gulf of Aden 54

h

Hawaii 59
  formation 34
  tsunami in 47
  volcano in 45
hazard maps 63–4
health spas 41
Hess, Harry 24
Hildoceras ammonites 9
Himalayas:
  effect on climate 61
  formation 5, 36, 37
  growth 37–8
Holmes, Arthur 23
Homo sapiens sapiens 9
horst 36
hot spots 34, 35, 66
  volcanoes on 52, 56, 59
hot springs 40, 41, 42, 57
Hutton, James 6
hydrothermal vents 41, 42, 66

i

Iceland, Mid–Atlantic Ridge in 23, 33, 52–3
igneous rock 16, 29, 66
  cycle 17
  dating 39

formation 52
index fossils 9
Indian Ocean, ridge in 22
Indian subcontinent 5, 37
iron 15
  magnetized 18
islands:
  deep–ocean 24–5
  volcanic 32, 34, 35
  volcanic chains 34, 35, 44, 59
isotopes 39

j

Japan:
  earthquakes in 47, 49
  tsunami in 47
  volcanoes in 62
jet stream 61
Jupiter 43
Jurassic period 9

l

lahars 62
Laurasia 11
lava 16, 50, 51, 65, 66
  domes of 58
  forming ocean floor 54
  molten 54–5
  rivers of 53
  speed of flow 52, 53, 54, 55
  types 52, 54, 55, 57
lead ore 52
limestone 8–9, 29, 55
lithification 17
lithosphere 15, 30, 51, 59, 66
Lyell, Sir Charles 6

m

magma 17, 33, 66
  chemical composition 52, 55–8
  viscosity 58
  from volcanic eruption 45, 51, 52
magnetic field 4, 18–20, 24
magnetic mineral 66
magnetic poles 20
magnetic reversal 18, 20, 25, 66–7
magnetic timescale 24

magnetometers 19, 20
magnetosphere 18
mantle 14, 15, 16, 67
  convection 26, 27, 65
  fluid layer 16–17
  plate movement relative to 34–5
mantle plume 34, 56, 59, 67
Marina district 64
Marianas Trench 21–2
Mars 43
marsupials 11
Matthews, Drummond 25
Mercalli, Giuseppe 49
Mercalli scale 48, 49
metamorphic rock 16, 29, 67
  cycle 17
  dating 39
metamorphism 17
meteorites 61
Micraster echinods 9
Mid–Atlantic Ridge 21, 22, 23, 33, 52–3
mid–ocean ridges 22, 23, 25, 26, 30, 67
  earthquakes on 47
  volcanoes on 51–4
mid–ocean rifts 33
minerals 41, 67
  dominant 15
Mohorovičič, Andrija 14
molten iron 15
molten lava 54–5
molten rock 16
  cycle 17
  eruption 26
monsoons 61
Moon 12–13
mountains:
  collision belts 30, 36
  extending into mantle 17
  formation 30, 32, 36, 37–8, 38, 44
  submarine 22, 23, 26
movement of tectonic plates 30
  direction 34
  relative to core and mantle 34–5
  relative to each other 31–4
mudslides 62
mudstone 8–9

n

Nanga Parbat 38
New Madrid, Mississippi,
  earthquakes 50
Northern lights 19
Nyiragongo volcano 54

o

ocean floors 16
  ageing 24, 25
  dating 23–4
  depths 21
  exploration by submersible
    40, 53
  formation 38, 54
  lava forming 54
  mapping 22, 31
  measuring from satellites
    31
  movement 25, 26, 31, 33,
    37
  organisms living on 40–2
  shape 21
  sinking 24
  trenches in 22, 26
oceans:
  formation 25, 31, 44
  islands in 24–5
  mountains in 22, 23, 26
  seamounts in 24, 67
  vanished 38
  volcanoes in 24, 52–4
On the Origin of Species 6
Ordovician period 8
oxygen 58

p

Pacific Ocean 13
  depths 21–2
  movement 34
  Ring of Fire 51
  volcanoes 51, 58
Pacific plate 55
  movement 34–5
palaeomagnetism 18, 67
  of ocean rocks 24
Pangaea 11, 39
Paradoxides trilobites 8
Pentacrinites crinoids 9
peridotite 15
Permian period 8
Pinatubo, Mount,
  eruption 61

planets, plate tectonics
  on 43
plate tectonics 28, 30–9,
  65
  on planets 43
plates see tectonic plates
polarization, reversing 18
Pompeii 60
Principles of Geology 6
productid brachiopods 9
pyroclastic flows 62

q

quartz 16
Quaternary period 9

r

radioactive isotopes 39
radioactivity 4, 7–10, 67
radiometric dating 7–10,
  19, 23, 67
Ranier, Mount 64
Red Sea 36
remote sensing 67
reversing polarization 18
Richter, Charles F. 49
Richter scale 49
rift valleys 22, 25, 30, 33,
  53, 67
  volcanoes in 54–5
rifting 36
Ring of Fire 51
rocks:
  carved by ice 10
  cycle 17
  dating 7–10
  dominant 15
  seismic waves travelling
    through 14, 15
  strata 6, 8–9
Roman Empire 60
Rwanda 54

s

St. Helens, Mount:
  volcanic eruption 58, 60
San Andreas fault 30, 34,
  47, 61
San Francisco: earthquakes
  48, 64
sandstone 8–9, 29
satellite images:
  information on plate

  theory from 42
  measuring ocean floors
    31
seamounts 24, 67
Seattle 64
sediment 17, 67
sedimentary rocks 8–9, 67
  cycle 17
  erosion 16
  fossils in 38
  layers 17
seismic tomography 43
seismic waves 14, 15, 43,
  67
seismology 46, 67
seismometers 14, 46, 67
shale 8–9
shrimp 40, 41
silica 58
Silurian period 8
solar wind 18–19
South America, coastline
  matching Africa 12, 13, 30
spas 41
spirifid brachiopods 8
starfish 40
Steno, Nicolaus 8
Stoffel, Dorothy 58
strata 6, 8–9
stratigraphy 67
subduction 67
subduction zones 30, 33, 38,
  47, 52, 55, 67
  oceanic 59
  volcanoes in 55–8, 59
Sullivan, Walter 65
sulphides 41, 52

t

tabulate corals 8
Tambora, Indonesia,
  volcanic eruption 60–1
tectonic plates 28, 67
  boundaries 32–3
  direction of movement
    34
  earthquakes on 46–7
  geometry 35–7
  movement 30
  on planets 43
  relative movement 31–5
  satellite information on
    42–3

speed of movement 31–4
Tennyson, Alfred, Lord 38
Tertiary period 9
Tethys Ocean 38
Thomson, William (Lord Kelvin) 7
thrusts 37
Tibetan plateau, effect on climate 61
Tokyo, earthquake 47
Tonga 59
transform faults 32, 33, 34, 47, 61, 67
tree rings 61
trenches 22, 26
Triassic period 9
trilobites 8
tropical plants: fossils 10, 28
tsunamis 47, 59
tube worms 40, 41
Turritella gastropods 9

u

uniformitarianism 6, 12
United States: movement 31–4
Unzen, Mount 62
uplift 16

V

Van Allen belts 18
vents:
  hydrothermal 41, 42, 66
  underwater 41, 42
Venus 43
Vesuvius 60
Vine, Frederick 25
Virginia opossum 11
volcanoes 51–9, 67
  active 51
  distribution pattern 28
  earthquakes linked to 46, 49, 60
  eruptions 6, 44, 45, 56–7
    causes 51–2
    impact of 60–1
    predicting 44, 62–3
  extraterrestrial 43
  fire fountain 45, 56
  fissure eruptions 57
  hot–spot 52, 56, 59
  inactive 59
  island–arc 59
  mudslides 62
  remains of 38
  rift–valley 54–5
  sea–floor 34, 59
  shapes 57
  subduction–zone 55–8

  types 57
  undersea 24, 52–4, 67
volcanic islands 24
  assessing plate movement from 34, 35
  formation 32, 34, 35
volcanic island chains 59
  formation 34, 35, 44

W

waves, action of 6
weather:
  affected by volcanic eruptions 60–1
  effects of 6
weathering 17
Wegener, Alfred 10–12, 13, 23, 30, 39
Wilson, John Tuzo 26, 34
Worldwide Standardized Seismometer Network 46

y

Yellowstone National Park 41, 60

z

Zebrowski, Ernest 62
zircon 39

further reading

The Dating Game: One Mans's Search for the Age of the Earth, Cherry Lewis, Cambridge University Press. ISBN: 0521790514

Target Earth: How Rogue Asteroids and Doomsday Comets Threaten our Planet, Duncan Steel, Quarto, 2000. ISBN: 070543365

Volcanoes Crucibles of Change, R. V. Fischer, G. Heiken and J. Hunter, Princeton University Press, 1997. ISBN: 0691002495

Perils of a Restless Planet, Ernest Zebrowski Jr., Cambridge University Press, 1997. ISBN: 0521573742

Why the Earth Quakes: The Story of Earthquakes and Volcanoes, Matthys Levy and Mario Salvadori, W. W. Norton and Co. ISBN: 0393037746

The Eyewitness Handbook of Fossils, Cyril Walker and David Ward, Dorling Kindersley, 1992. ISBN: 0751310042

Volcanoes, R. Decker and B. Decker, W H Freemans and Co. (including CD Rom). ISBN: 0716731746

On the Rocks Earth Sciences for Everyone, John S. Dickey Jr., John Wiley, 1988. ISBN: 0471132349

Continents in Motion: the New Earth Debate, Walter Sullivan, AIP, 1991. ISBN: 0883187043

....

acknowledgments

Jacket design: Nathalie Godwin
Proof-reading: Jane Simmonds

picture credits

The publisher would like to thank the following for their kind permission to reproduce their photographs. KEY: SPL = Science Photo Library.

1: SPL; 4 background: NASA/JPL/SPL; 5: D Wilman/Art Directors/Trip; 6: Dr. Jeremy Burgess/SPL; 9: Jim Amos/SPL; 11 top: SPL; 11 below: Joe McDonald/Corbis; 20: Amos Nachoum/Corbis; 21: Mary Evans Picture Library; 22: Institute of Oceanographic Sciences/NERC/SPL; 23 below: SPL; 24: Department of Earth Sciences, Princeton University; 31: Robert Harding Picture Library; 32 below right: David Parker/SPL; 33 top: Bernhard Edmaier/SPL; 33 below: David Parker/SPL; 34: Archives of the University of Toronto Library; 36 background: Helene Rogers/Art Directors/Trip; 37 background: NASA/SPL; 40: Stan Wayman/SPL; 41 top: Dr Ken Macdonald/SPL; 41 below: Tony Craddock/SPL; 42: David Hay Jones/SPL; 45: Krafft/Explorer/SPL; 47 below: The Art Archive/Claude Debussy Centre St. Germain en Laye/Dagli Orti; 49 top: Bettmann/Corbis; 49 below: Istituto Nazionale di Geofisica e Vulcanologia, Italy; 50: Used by Permission of the State Historical Society of Missouri, Columbia; 53 top: Stuart Westmorland/Getty Images; 53 below: Southampton Oceanography Centre UK; 54 top: SPL; 54 below: EPA/PA Photos; 55: Dr. Morley Read/SPL; 56: Bernhard Edmaier/SPL; 58: Roger Ressmeyer/Corbis; 60: Museo Archeologico Nazionale di Napoli/Dorling Kindersley; 61: Royal Tyrell Museum, Canada/Dorling Kindersley; 62: Roger Ressmeyer/Corbis; 63: Jeremy Bishop/SPL; 64 top: Peter Menzel/SPL; 64 below: George Olson/SPL; 65: Roger Ressmeyer/Corbis.

All other images © Dorling Kindersley. For further information see: **www.dkimages.com**